Chrispin Yekia Bayolo
Jean Paul Ngbolua Koto-Te-Nyiwa
Ruphin Djolu Djoza

Contribution to the biodiversity inventory at Jardin de Gbadolite

Chrispin Yekia Bayolo
Jean Paul Ngbolua Koto-Te-Nyiwa
Ruphin Djolu Djoza

Contribution to the biodiversity inventory at Jardin de Gbadolite

Biodiversity status and prospects at the Jardin botanique et zoologique de Gbadolite (Nord-Ubangi, DR Congo)

Imprint

Any brand names and product names mentioned in this book are subject to trademark, brand or patent protection and are trademarks or registered trademarks of their respective holders. The use of brand names, product names, common names, trade names, product descriptions etc. even without a particular marking in this work is in no way to be construed to mean that such names may be regarded as unrestricted in respect of trademark and brand protection legislation and could thus be used by anyone.

Cover image: www.ingimage.com

This book is a translation from the original published under ISBN 978-620-6-71256-5.

Publisher:
Sciencia Scripts
is a trademark of
Dodo Books Indian Ocean Ltd. and OmniScriptum S.R.L publishing group

120 High Road, East Finchley, London, N2 9ED, United Kingdom
Str. Armeneasca 28/1, office 1, Chisinau MD-2012, Republic of Moldova, Europe
Printed at: see last page
ISBN: 978-620-7-76632-1

EPIGRAPH

A termite mound attached to a shrub will not fall off.

"YEKIA BAYOLO Chrispin

DEDICATION

To you, dear Papa Apollinaire NDANGA BIKIBO and Mama la Rose TOLOLI, for your invaluable support and encouragement. Thanks to you, we have become what we are today, and we will always be very grateful.

"Chrispin YEKIA BAYOLO

ACKNOWLEDGEMENTS

We would like to express our sincere thanks to all those who have contributed in any way to the production of this work.

We would like to take this opportunity to thank the academic authorities and teaching staff of the University of Gbadolite (UNIGBA) for the know-how and expertise they have made available to us.

Our gratitude goes to Professor Dr. *Jean Paul NGBOLUA KOTO-TE-NYIWA* for directing this work;

We'll never forget Assistant *Rufin DJOLU* for his pertinent comments and guidance.

Our warmest thanks also go to all the academic, scientific and administrative staff of our Faculty, and in particular those of the Environmental Sciences Department, notably Prof. Jeff ITEKU BEKOMO, Alain KAMIKA, GONZATO BINA. The supervisors NZANGBA-DU-LUTA, Maurice NGEMALE, John LIKOLO BAYA, the assistants Amédée GBATEYA, Michau KAMIENGE, Armand ENDOWA, Honoré YABUDA and Sammy NGUNDE. We have never hesitated to call on one or other of them for any problem, no matter how small. Their advice and answers have always brought us relief and comfort. May they accept our deepest gratitude.

To our parents *Gaspard EKULU LILONGO* and *José BOKUMBE BAYOLO* for what they have always been to us;

Our thanks also go to Pastor *Fils BOKESE ENGO* of **La Borne Gbadolite** church, for his spiritual guidance and advice. Far or near, we'll always be together.

To our brothers, sisters and nieces: Ange Fidel ENGUNGA, IYENDE LOMBOTO, Dady BAYOLO, Alfred BAYOLO, Nathan BAYOLO, Dido

LIKOMBO, Techa EWUZU, Tonton EWUZU, Dieu Merci BASELE, Nicolas MATANGO, Dieu Merci SAMBA, Sandra BAYOLO, Claudine BOLUMBU, Landrine EKILINGA, ELODI, Sarah IKOMBA, Jemima LIANYA , Nadine IMPAMA, Adel YAGANZA, Bienvenu OMOLITE and Sita MATEMBE For their brotherly love.

To Docteur. Jaques, Joël LUEMBA, Jérémie BONTAMBA, Dr. Héritier MBETE, Rigoberd BOLAMPANZA, Alois LIKONGO, Nono NSUMBU, Tathy BIKAMBA, IYOKA TOZIRO, Calvin KOYEMBA, Gérard BASENGE, Elie BAMONGA, Patrick ELOBO, Frédérique MPIA, Jean AKONA for your precious support;

 To our valiant comrades in arms: Faustin MOTENDO, Giscard YANGU, Chadrack BAKWA, Arsène BOONGA, Patrick ILANGA, Olivier MAKILI, Olive LOMO, José IFAMBE, Roger BONZE, BOPALO ELEZA, Cédric TUBU, Sandra BOKOTSHI, Déborah MITONGO, Jupsie BOLUNGA, for their friendship.

To all those who have not been mentioned by name, but whose moral support and presence at our side have made a great contribution to the realization of this work, may they find here the expression of our sincere gratitude.

"Chrispin YEKIA BAYOLO

0. INRODUCTION

0.1. Issues

Today, biodiversity issues are more important than ever. This is the context in which biological invasions are taking place. Many people, whether involved in the environment or not, believe that these phenomena can only enrich biodiversity through the arrival of new species and thus increase species richness. Unfortunately, numerous scientific studies on exotic species have shown that some of them are invasive (causing damage to the environment). Indeed, the arrival of a new species in an ecosystem inevitably leads to interactions with the newly colonized environment, and therefore with the other species present. These interactions can have a detrimental effect on ecosystem functionality, and it is only the complexity and multitude of ecosystem functionalities that define its richness, and therefore its biodiversity. (Aulagnier, 2008; Babski, 2009; Baker, 2008; Barbault, 2010).

Two of the main missions of the National Species Conservation Department are to understand and protect exotic species in natural and semi-natural habitats. To this end, exotic species are evaluated and studied by monitoring their protection, distribution and behavior within the Garden, in order to identify those species that are problematic in terms of natural heritage, species conservation and the need for control measures (Aboucaya, 1999).

One of the main difficulties at present is the lack of knowledge about the benefits and real impact of exotic plants on biodiversity in the D.R. Congo, especially at the Jardin botanique et zoologique de Gbadolite. Studies in this area are therefore recommended. The Jardin botanique et zoologique de Gbadolite was founded over 45 years ago. However, between 1985 and 1990, the Garden was known as the "Ubangi Green Space and Zoo".

In 1997, the garden was destroyed following the rebellion and abandoned until 2005.

In 2008, the Jardin botanique et zoologique de Gbadolite was integrated into the Institut des Jardins zoologiques et botaniques du Congo by Ministerial Order N°017/CAB/MIN/ECNT/2008 of July 17, 2008.

The present study was initiated with the aim of making an inventory of the animal and plant species conserved ex situ in this Garden, with a view to proposing new perspectives for the sustainable management of the exceptional biodiversity of this heritage.

0.2. Choice and interest of subject

The present work is of threefold interest:

- ➢ **On the environmental front:** data from this research will serve as an effective tool for the rehabilitation project of the Gbado-Lite Botanical and Zoological Garden in its role of conserving biodiversity and serving as an ecological framework for natural resources. In fact, this garden will still enable the population to learn about ex-situ exotic animal and plant species, a profound and very important culture.

- ➢ **Economically:** the present work contributes to the evaluation of exotic species in the Botanical and Zoological Garden of Gbado-lite. The latter makes a major contribution to the economy of the town of Gbado-lite, considering animals and plants as a source of funding. In fact, researchers buy them simply by observing them, which is what tourism is all about.

- ➢ **Scientific:** The data from this study will help scientists and the general public to manage exotic species sustainably in protected areas and the botanical and zoological garden in general, and will serve as a reference document for agents working in this field in particular.

0.3. Division of labor

Apart from the introduction and conclusion, the present work comprises three chapters:

- ➢ The first deals with General information on protected areas;
- ➢ The second concerns Environment, Materials and Methods;
- ➢ The last section deals with Results and Discussion.

CHAPTER I. GENERAL

I.1. PROTECTED AREAS

I. 1. 1 Definition

A protected area is a geographically clearly defined zone, delimited, regulated, recognized, dedicated and managed by any effective means, in order to ensure the long-term conservation of nature, ecosystem services and associated cultural values (Ferrez, 2006).

Protected area is a term used to designate territories of varying size and conservation status. In descending order of importance of protection measures, there are six categories of such areas: strict nature reserves, national parks and monuments, specialized reserves, protected landscape zones, and resource management areas. These IUCN categories are complemented by two types of protected area created by UNESCO: biosphere reserves and world heritage sites (Ramade, 2008).

The success of protected areas as a conservation tool is based on the idea that they are managed to protect the values they contain. To be effective, management must be tailored to the particular demands of each site, given that each protected area presents a variety of biological and social characteristics, pressures and uses. Achieving effective management is not an easy task - it requires the adoption of appropriate management objectives and governance systems, adequate and sufficient funding, and the timely use of site-specific management strategies and processes (Musibono, 2012).

According to Law N°14/003 of February 11, 2014 on the management of natural resource conservation in the Democratic Republic of Congo, a protected area is defined as a clearly defined geographical space, recognized, enshrined and managed by any effective means, legal or otherwise, in order to ensure the long-term conservation of nature as well as the ecosystem services and cultural values associated with it (Musibono 2013).

I.1.2. Types of protected areas in the DRC
I.1.2.1 Reserves

Depending on the degree of human intervention and accessibility for visitors, there are 4 types of nature reserves (Holway, 2007):

- **Forest reserves:** are forests or parts of forests classified in accordance with current legislation, with the aim of safeguarding characteristic or remarkable facies of native species stands and ensuring the integrity of the soil and environment (Monolou, 2001).

- **Strict nature reserve or scientific reserve: this is a** category of protected areas set aside to protect biological diversity and possibly also geological and/or geomorphological features, where visits, use and human impacts are strictly controlled and limited to guarantee the protection of conservation values. Access is restricted to government officials and qualified scientists. This reserve offers a number of advantages, as the absence of human intervention leaves ecosystems free to evolve in a positive direction. The ponds can be seen to be reconstituted, filling in with grasslands that provide food for the benthic population; (GIRAUD, 2001).

- **Directed nature reserve**: this is a protected area established for the purpose of scientifically monitoring and guiding the evolution of nature. In principle, access is restricted to certain persons responsible for scientific or administrative supervision. The management of controlled nature reserves normally involves scientific intervention to help maintain semi-natural biocenosis, and the monitoring of certain species threatened by competing species (Ruys, 2009).

- **Strictly wilderness reserve: this** is an area whose main purpose is to protect the natural environment:

    Protecting nature;

    To allow a descent for visitors able to bear the comforts of visiting an area maintained strictly in its wild state. That's why this area has no form of transport, no campgrounds, no development, no authorized transport and a regularly maintained trail. So everything must remain in its pristine state, and the free existence of native plant and animal species is fully assured (Parap, 2013).

I.1.2.2. Parks

It is a territory maintained in a natural or semi-natural state, for conservation or relative purposes. (Sri paul, 1965).

It is a reserve of land, most often, but not always, declared and determined by a government away from most human and national development to be a UICCN Category II protected area (Ulyssess, 1872).

I.1.2.3. Gardens

The Gardens of *ex-situ* and *in-situ* protection strategies, in which mankind constitutes a sort of modern-day Noah's Ark that can only host a small proportion of biodiversity (Anonymous, 2008).

The practice of creating gardens to acclimatize plant and animal species began in Europe as early as the $18^{\text{ème}}$ century.

Since the beginning of the present century, it has played an important role in the preservation of plant species threatened with extinction in their original ecosystems. A botanical garden is an area set up by a public institution or association with the aim of bringing together documented collections of living plants for conservation, scientific research, exhibition, tourism or teaching purposes (Giraud, 2001).

The zoological garden or *zoologicalgarden* or simply Zoo is a space where animals of wild species or exotic domestic species are maintained and raised in captivity for the purposes of conservation, scientific research, exhibition, tourism or teaching. (Mandjo 2014).

CHAPTER II. ENVIRONMENT, MATERIALS AND METHODS

II. 1 Study environment s

The species data were collected at the Jardin botanique et zoologique de Gbadolite, located in the province of Nord-Ubangi, three kilometers from the town of Gbadolite, and authenticated by assistant Honoré YABUDA, researcher at the Faculty of Agronomic Sciences of the University of Gbadolite.

- **GPS COORDINATES**

The Jardin Botanique et Zoologique de Gbadolite(JBZG) covers 300 hectares and is perched at an altitude of around 405 m above sea level. It is located at latitude 4°14' North and longitude 21°4' East, 3 km from the center of the town of Gbadolite in the New Province of Nord-Ubangi.

II. 2. Geographical location
II.2.1 City of Gbadolite

The town of Gbadolite is the capital of the New Province of North Ubangi. It is located at 21° 1' longitude East and 4° 17' latitude North.

II.2.2. Gbadolite Botanical and Zoological Garden

➢ **Delimitation**

The limits are defined :

- **North**: Former official presidential residence separated from JBZG by the coconut grove;

- **North-West**: The bed of the Wakamba river at the junction with the BOYI river;

- **To the north-east**: the public cemetery and the CDAI sprl palm grove;

- **To the south-east:** the CDAI sprl palm grove and the JBZG fish ponds;

- **Southwest**: The Wakamba riverbed and the MANGUNDU road.

> **GENERAL LAND USE**

The concession has a surface area of 300 hectares, subdivided into 6 blocks along 4 main axes and invested respectively with the following flagship collections:

. Block A : 20 Hectares

- WAKAMBA River
- Leceuna
- Begonias
- Nursery
- Young arboretum

- Cocoa
- Heliconia
- Storage and rest shed
- Salute to the Flag

. Block B : 15 Hectares

- Visibility wall
- Black wood(Eben)
- Papayeraie
- Hura crepitans
- Devil's tree
- Banana plantation

- Palmetto
- Yellow bamboo
- Animal park
- Fishpond
- Papyrus

. Block C: 15 hectares

- Administration building
- Young orchard
- Bergerie
- Fields

. Block D2: 150 hectares

- Mango trees
- Safoutier
- Fields

- **Living botanical collections**

. Block D1: 40 hectares

- Ornamental park
- Mango trees
- Fields

● Block E: 60 hectares

- Cocoteraie
- Masonry foundation for the destroyed biotopes of the former ZOO
- Masonry foundation of destroyed residential houses
- Fields: former log orchard, coffee and avocado trees

Timber and non-timber forest products (NTFP) identified and collected are classified as follows: **78 orders, 93 families, 309 species and 3476 specimens**.

The botanical collections are characterized by :

- The predominance of endemic and exotic ornamental plants,
- 64% medicinal botanical collections,
- Mango, coconut and safoutier plantations,
- *Cyperus papyrus* introduced in 1980,
- Fructification of *Couroupita guinensis,* known as the Devil's Tree and preferred host of edible caterpillars,
- Young arboretum of timber species,
- Young Safoutier orchard under development

II.3 Materials

The materials used in our research are as follows:

- ❖ Notepad and pen,
- ❖ Laptop computer,
- ❖ Digital camera.
- ❖ GPS.

II.4 Methodologies

This project involved a systematic inventory of the biodiversity conserved ex-situ at the Gbadolite botanical and zoological garden.

CHAPTER III. RESULTS AND DISCUSSION.

II.1 Plant taxa of the Jardin botanique et zoologique de Gbadolite

Figure 1 shows the number of horticultural species at the Jardin Botanique et Zoologique de Gbadolite before and after the war.

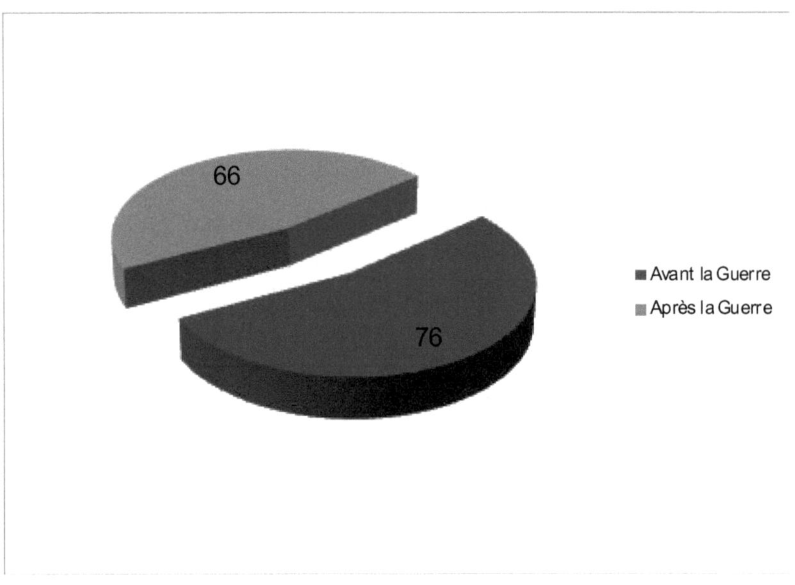

Figure 1: Number of horticultural species before and after the war

This figure shows that the Jardin Botanique Zoologique de Gbadolite was home to 76 horticultural species before the war. Today, the Garden has only 66 of these species, meaning that 10, or 13.15%, have disappeared.

Figure 2 shows images of some of the horticultural species found in the Gbadolite JBCZ.

Figure 2 *(a) Alternathera correae*

(a)

Figure 2 b shows images of the *Aloe vera* nursery at the Jardin Botanique et Zoologique de Gbadolite.

(b) Aloe vera nursery

(c) Anthurium correae nursery

Figure 3 shows the exotic forest species at the Jardin Botanique et Zoologique de Gbadolite before and after the war.

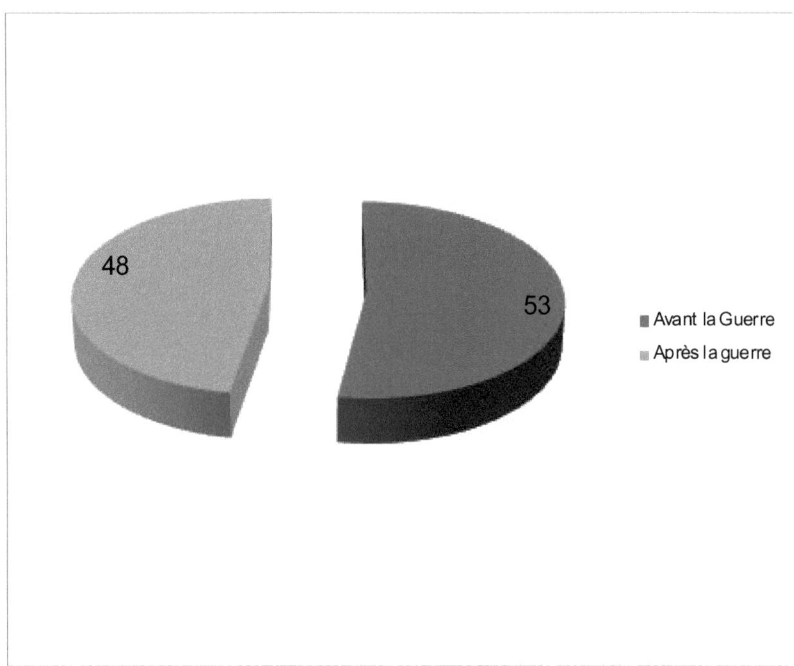

Figure 3 Number of forest species before and after the war.

This figure shows that the Garden had 53 species. However, after the war, only 48 species remained, a loss of 5 forest species representing 9.43%. These species are : *Chrysophylium africana, Entandrophragma angolense, Entandrophragma cylindricum, Funtumia africana, Funtumia elastic.*

Figure 4 shows images of two flagship species

Figure 4 (a) Albizia ferragena

Figure 4 (b)Couroupita guianensis

Table 1 lists forest plants.

N°	Plant species	Common name	Number of existing species	
			Before the war	After the war

No.	Species	Local name		
1.	*Albizia ferragena*		+	+
2.	*Albizia lebbec*		+	+
3.	*Annona reticulata*		+	+
4.	*Anonidium mannii*	*Endenge*	+	+
5.	*Boerovia diffusa*		+	+
6.	*Bacteria ecydiya*		+	+
7.	*Bombax buonopozense*	*Kapokier*	+	+
8.	*Cassuessus angolensis*		+	+
9.	*Ceiba pentandra*	*Fuma*	+	+
10.	*Chrysophyllum africana*	*Longii white*	+	-
11.	*Coffea canephora*	*Bohaka*	+	+
12.	*Cola acuminata*	*nkuzu*	+	+
13.	*Cola nitida*	*Colani*	+	+
14.	*Couroupita guianensis*		+	+
15.	*Dacryod esedulis*	*Bosau*	+	+
16.	*Erypheleum guianensis*		+	+
17.	*Dacryodes pubescens*	*Bosaw*	+	-
18.	*Entandrophragma angolense*	*Tiama*	+	-
19.	*Entandrophragma candollei*	*Ngola*	+	+
20.	*Entandrophragma cylindricum*	*Lifakepembe*	+	-
21.	*Useful Entandrophragma*	*Kote*	+	+
22.	*Heveabras iliensis*	*Hevea*	+	+
23.	*Hylodendronga bunensis*	*Pangu*	+	+

			+	+
24.	*Irvingia gabanensis*	*Bofalanga*	+	+
25.	*Eucalyptus agave*		+	+
26.	*Euphorbia huirti*		+	+
27.	*Ficus elastica*	*Lowa*	+	+
28.	*Funtumia africana*	*Mutondo*	+	-
29.	*Funtumia elastica*	*Wembe-bo-kikeleke*	+	-
30.	*Garcinia punctata*	*Bosefe*	+	+
31.	*Ixora ordorata*	*Angwondia*	+	+
32.	*Khaya grandifolia*	*African mahogany*	+	+
33.	*Makaranga laurentii*	*Lokoke*	+	+
34.	*Milicia exelsa*	*Kambala*	+	+
35.	*Millettia laurenti*	*Wenge*	+	+
36.	*Paciflora angolensis*		+	+
37.	*Parkia bicolor*	*Kungulongo*	+	+
38.	*Pentacletra macrophilla*		+	+
39.	*Pterocarpus soyauxil*	*Bosulu*	+	+
40.	*Philodendro déwevrei*		+	+
41.	*Pycnantus angolensis*	*Ilomba*	+	+
42.	*Pycranthus angolensis*		+	+
43.	*Jatropha congolensis*		+	+
44.	*Rowolfia vomitoria*		+	+
45.	*Ricinodendron hendelotii*	*Bofeko*	+	+
46.	*Ricinodendron communis*		+	+

47.	*Schrebera gabonensis*	*Kiala*	+	+
48.	*Tectonia grandis*		+	+
49.	*Terminaliasuperba*		+	+
50.	*Vernonia brazzavillensis*		+	+
51.	*Uapaca guianensis*	*Rikio*	+	+
52.	*Zanthroxylum gilletil*	*Olonvogo*	+	+
53.	*Zanthroxyluml aurentii*	*Pangi*	+	
	TOTAL		53	48

The table above shows that of the 53 forest species conserved at the Gbadolite JBZ, 48 are still alive today and 5 are extinct.

Table 2 lists the fruit species conserved at the Gbadolite JBZ.

N°	Plant species	Common name
1.	*Anoniduim mannii*	
2.	*Annona mystica*	
3.	*Artocarpus integrifolio*	
4.	*Artocarpus integrifolio*	Canistel
5.	*Carica papaya*	Papaya
6.	*Citrus limon*	
7.	*Cocos nucifera*	
8.	*Flacutiyaj angomas*	Jam
9.	*Dacryo e edilis*	Safutier
10.	*Elaeis guineensis*	
11.	*Eugenia ham*	
12.	*Flacourtia jangomos*	
13.	*ZingebeR oficinal*	
14.	*Mangifera indica*	Mango tree
15.	*Roystonéa regia*	
16.	*Persea americana*	Avocado tree
17.	*Psidium guajava*	Guava
18.	*Teobroma cacao*	Cocoa
TOTAL		18

It should be noted that only fruit species have been preserved to date at the Jardin Botanique Zoologique de Gbadolite.

Figure 5 shows the fruit species.

Figure 5 (a)*Mangifera indica*

(b) *Dacryodes edulis*

II.2 Animal taxa

II.2.1. Hoofed mammals

Before the war, two hoofed mammals were kept ex-situ at the JBZ EE Gbadolite. These included *Elephas maximu* (Mastrondadronte) and *Elephas maximus sumatranus* (Manaute 6-horned elephant).

Figure 6 (a) *Elephas maximus* (b)*Elephamaximus sumatranus*

These animals no longer exist at the Jardin Botanique et Zoologique de Gbadolite. Image (Géorge Cuvier 1998).

III.2.2 Carnivorous mammals

Table 3 List of carnivorous mammals preserved at the Gbadolite e before and after the war.

N°	Animal species	Common name
01	*Canis aurens* (Fréderic knutter 1940)	Jackal
02	*Dologale dybowskii*(Sri paul Edward 1950)	Dybowski's mongoose
03	*Herpetes ichneuon (Cuvier 1798)*	Egyptian mongoose
04	*Hrpeste urva* (Cuvier 1798)	Crab mongoose

05	*Ursus actos* (Ulysses, 1872).	The Brown Bear
06	*Vulpes vulpes (*Maupass, 1988)	Fox
07	*Taurotragus oryx (*Saut, 1885)	Cape Eland
08	*Acrocodia indica* (Tales, 1911)	Rufous tapir
10	*Giraffa camelopardalis* (Atrementowiecz, *2010)*	Giraffe
11	*Ammostragus levia(*Saut, 1885)	Mittens
12	*Robust Sapajou* (Frédericknutter, 1940)	Alaskan monkey
13	*Callicebu sornatus*	Brazza monkey
	TOTAL	13

This table shows that 13 species of carnivorous mammals were at the Jardin Botanique et Zoologique de Gbadolite before the war, 11 have disappeared, and only two species remain today.

These include *Callicebus ornatus and Sapajou robuste.*

Figure 7 shows the species of carnivorous mammals.

Figure 7 (a) Callicebus ornatus

(a)

Figure 7 (b) Robust Sapajou

(a)

Figure 4: Ungulate mammal species conserved at the Jardin Botanique et Zoologique de Gbadolite before and after the war.

N°	Animal species	Common name
01	*Commus squirrels (Barbault 2006)*	4-banded squirrel 20 cm
	TOTAL	1

Figure 8 shows the ungulate mammal species

Figure 8 *Commus squirrels*

Before the war, one species of ungulate mammal was kept ex-situ at the JBZ Gbadolite. After the war, this species disappeared.

Table 5: List of reptiles kept at the Jardin Botanique et Zoologique de Gbadolite before and after the war.

N°	Animal species	Common name
01	*Crocodilus nilaticuslaurenti*) (Chaldea, 1964)	Crocodile sial mois
	TOTAL	1

This table shows that before the war, the ordinary reptile species existed in the Jardin Botanique et Zoologique de Gbadolite, and is still alive today.

Figure 9 shows the common reptile species

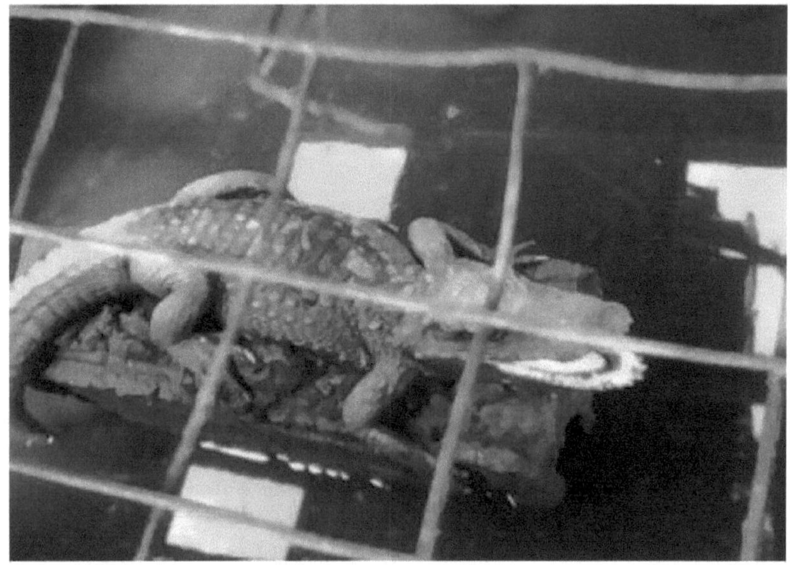

Figure 9 Crocodilus laticuslaurenti

Table 6: List of reptiles and amphibians preserved at the Jardin Botanique et
Zoologique de Gbadolite before and after the war.

N°	Animal species	Common name
01	*Chelodina longicollis*	Oriental box turtle
02	*Boa constrictor*	Boa constrictor
03	*Crotalus scutulatus*	Rattlesnake
04	*Chrysopelea indonesia*	Flying snake
	TOTAL	04

The table shows that the Jardin Botanique et Zoologique de
Gbadolite once had 4 species of amphibian reptiles, none of which exist
today. All have been lost to the war.

Table 7: List of endangered animal species in the Gbadolite JBZ before and
after the war.

N°	Animal species	Common name
01	*Chelonia mydas*	Green turtle
02	*Varanus komodaensis*	The biggest lizard
03	*Loris tardigradus*	Hail sow
	TOTAL	

This table shows that 3 species of endangered animals were at
the Jardin Botanique et Zoologique de Gbadolite before the war, but all
disappeared after the war.

Table 8: List of poultry species at the Jardin Botanique et Zoologique de Gbadolite before and after the war.

N°	Animal species	Common name
01	*Anser anser*	Khaki goose fill
02	*Anser fabalis*	Egyptian goose
03	*Cuculus canorus*	Black Antwerp Cuckoo
04	*Cuculus linnaeus*	Cuckoo from Antwerp thousand
05	*Anser indicus*	Bar-headed goose
06	*Anos platyntynchos*	Peking duck
07	*Necrosyrtes monochus*	Angola Vulture
08	*Musophagiformes*	Touraco
10	*Passer domesticus*	Moapa
	TOTAL	10

This table shows that 10 bird species were present at the Jardin Botanique et Zoologique de Gbadolite before the war, none of which still exist today. All disappeared after the war.

III.3. Outlook

The Jardin Botanique Zoologique de Gbadolite practises ex-situ conservation on a 300-hectare concession that is relatively visible as a result of daily activities that are gradually attracting visitors, an average of four people a month, mainly from students of all primary, secondary and higher education levels. Management constraints are considerable, and require the support of technical and financial partners in the following activities:

- Botanical propagation nursery,

- Planting and maintenance of botanical collections,

- Care for animals in semi-liberty,

- Breeding domestic exotic and wild species,

- Renewable energy production,

- Construction of conservation, reception and safety infrastructures,

- Supply of materials, equipment and various products.

In a nutshell, the Botanical and Zoological Garden is subject to a number of constraints, which are resolved by projects with different timeframes and rapid impact, ranging from the short term to the long term.

CONCLUSION AND SUGGESTIONS

This book looks at the biodiversity of the Jardin botanique et zoologique de Gbadolite before and after the war, the current situation and prospects for possible rehabilitation.

At the end of this study, we have shown that :

- The Gbadolite botanical and zoological garden had 3 types of exotic plant species (Horticultural, Forestry and Fruit) before the war, grouped into 147 species. Today, 132 species are present. So 15 species have disappeared because of the war. Apart from ornamental species, all fruit and forest species are in a state of ageing.

- As for exotic animal species, the botanical and zoological garden had 7 types of species (hoofed mammals, carnivorous mammals, ungulate mammals, reptiles, amphibian reptiles, endangered species and poultry animals) before the war, grouped into 34 species, and today 3 species are present. So 31 species have disappeared because of the war.

Taken together, these results constitute scientific evidence validating the merits of protecting exotic species for botanical and zoological gardens in the Democratic Republic of the Congo.

It would therefore be advisable to raise public awareness of the management of the Botanical and Zoological Gardens and the protection of biodiversity.

BIBIOGRAPHY

1. Aboucaya, A, 1999Bilan d'une enquête national destinée à identifier les xénophytes invasif sur le territoire métropolitain Français.
2. Ferrez Y, 2006. Definition of a strategy to combat invasive species in Franche-Comité, proposal for a hierarchical list.
3. D.E. Musibono, 2012. Toxic Policies and Pollution of the Global Environment: a collective suicide. Ed. ERGS-Kinshasa.
4. E. Musibono,
5. 2013 du marasme d'Etat squelette aux défis de développement durable-Gestion de l'Environnement au Congo-Kinshasa: Cueillette chronique et pauvreté durable. Ed. UNESCO Chair, Kinshasa.
6. B. Mandjo, 2014. Introduction to environmental sciences in D.R. Congo.
7. Barbault 2016; Atrementowicz 2010. Les invasions biologiques, une question de natures et de sociétés.
8. Giraud, 2001. Le Crime contre la nature. Paris, Ed. L'HARMATTAN.
9. Monolou J.C. 2001 Diversity. Biological dynamics and conservation.
10. Dictionnaire encyclopédique des sciences de la Nature et de la Biodiversité (Ramade, 2008),
11. PARAP; 2013; Programme d'Appui au Réseau des Aires Protégées, ICCN, DRC.
12. Baker R.H.A.,& al., 2008. The UK risk assessment scheme for all non-native species. In Rabitsch W., Essl F. &Klingenstein F. (Eds): Biological Invasions - from Ecology to Conservation. NEOBIOTA, 7 : 46-57.

13. Barbault R., 2016. Biological invasions, a question of natures and societies. Editions Quae. 180p.

14. Ruys, 2009. History and ecological characteristics of the invasion of the French Ardennes by three aquatic rodents: Muskrat (Ondatrazibethicus), Coypu (Myocastor coypus) and European Beaver (Castor fiber). Thesis defended December 18, 2009. University of Reims Champagne-Ardenne. Groupe d'Etude sur les Géo matériaux et Environnements Naturels/ Centre de Recherche et de Formation en Eco-éthologie (CERFE).

15. Sri paul, &Ovenden D., 1950. Le guide herpéto - 199 amphibiens et reptiles d'Europe. Delachaux et Niestlé, Paris, 288 p.

16. Géorge cuvier, Mandon-Dalger I., et Al, 1798. Protocoles de hiérarchisations biologiques invasives en vue de leur gestion.

17. Frédérique Saint Andrieux C., Pfaff E., Guibert B., 1940. Le daim et le cerf sika en France: nouvel inventaire. Faune et Flore Sauvage ONCFS n°285. p. 10-15.

18. Mauposs1988; Mbun knowledge and know-how in the sustainable conservation of biological ecosystems.

19. Saut, Lévêque C., 1985. Introducing species into aquatic environments. Should we be afraid of biological invasions? Editions Quae. 232 p.

20. Chaldée, JL., Marmet J., 1964. Ecureuils d'Europe occidentale: Fiches descriptives. 11p.

21. Atremento wiecz, Dutartre A., 2010. Managing biological invasions. ONEMA training course on invasive alien species in aquatic environments, April 12-16, 2010, Paraclet.

22. Conte, E Elton CS., 1911. The ecology of invasions by animals and plants, Chapmann and Hall, London.

23. Ulisses, 1872. much information on population biology is needed to Animal introduced species? Conservation biology 17: 83-92.

24. Giraud, 2001. Le Crime contre la nature. Paris, Ed. L'HARMATTAN.

25. Maupass, 1988 ; Savoir et savoir-faire Mbun dans la conservation durable des écosystèmes biologique.

26. Géorge Cuvier et al, 1998. Protocols for biological hierarchies with a view to their management.

TABLE OF CONTENTS

Printed by Books on Demand GmbH, Norderstedt / Germany